Reference Systems and Electrochemical Cells

By Malika Ammam, PhD

Copyright© 2017 Malika Ammam. All rights reserved.

Discount Offers

5% OFF of the book price for purchases of 1-5 books.

8% OFF of the book price for purchases of more than 5 books.

To receive the discount money, send your request through https://www.malika-ammam.com/ with your order details and PayPal account. Make sure that your order details (amazon or other sites) passed the 30 days return policy.

Thank you,

Introduction

As a teacher of physical chemistry, I noticed that students, even in advanced classes, have difficulties in understanding the basics of redox chemistry. The quantification of redox quantities, such as potentials requires reference systems (or electrodes) for comparison. Section 4 discusses the importance of reference systems (or electrodes) in the determination of electrochemical quantities. Several reference electrodes with their standard potentials are provided, and the trends of redox potentials across the periodic table are discussed. In addition to reference electrodes, the quantification of redox quantities requires electrochemical cells. Section 5 summarizes the basic components of electrochemical cells in terms of electrode and electrolyte, as well as the main mathematical quantities governing the redox reactions at equilibrium explained by the Nernst equation. To further clarify the discussed concepts, numerous questions and problems with detailed answers are provided. Most of these questions are formulated by students like you. I believe that these two Sections (4 and 5) would greatly help students with levels varying from high school to advanced university classes.

Section 4

Importance of Reference Systems in Determination of Redox Potentials

Basics Concepts with Resolved Questions and Problems

Abstract

The previous section provided a short concise definition of redox reactions and proposed ways to determine the oxidation states of elements, as well as how to balance redox reactions. This section discusses the importance of reference systems (or electrodes) in electrochemistry for determining the potentials of redox reactions at conditions different from the standard. The variation trends of redox potentials across the periodic table are also summarized.

1. **Reference systems or electrodes**

Reference systems or electrodes are characterized by well-known and stable potentials or voltages, which are obtained by buffered or saturated solutions of electrolytes containing redox participating species in the processes[1-7]. References electrodes are often used as half-cells in electrochemical cells to sense or determine the potentials of overall reactions. In theory, normal hydrogen electrode (abbreviated as, NHE, $2H^+ + 2e^- \leftrightarrow H_2$) is a perfect reference electrode, but cannot be employed for practical purposes due to risks of explosion if the hydrogen leaks out and reacts with the oxygen of the atmosphere. To overcome this safety problem, other reference electrodes are used instead of hydrogen electrode. Examples include saturated calomel electrode (SCE) based on the reaction between elemental mercury and mercury(I) chloride ($Hg_2Cl_{2(s)} + 2e^- \leftrightarrow 2Hg_{(l)} + 2Cl^-$), silver/silver chloride electrode (SSCE) involving redox reactions between silver metal (Ag) and its salt silver chloride AgCl ($AgCl(s) + e^- \leftrightarrow Ag(s) + Cl^-$), and copper-copper(II) sulfate electrode (CSC) implicating copper metal and its salt copper(II) sulfate ($Cu^{2+} + 2e^- \leftrightarrow Cu$).

2. **Standard potentials**

The tendency of chemical species to acquire or donate electrons is measured in terms of reduction or oxidation potentials, which are also linked to the ionization energies[1-8]. Thus, the measurement of potentials or energies is of great importance to evaluate the feasibility of redox processes under particular conditions. At the standard conditions (temperature 298 K or 25 °C, pressure 1atm, concentration 1 mol L^{-1}), the potentials of redox reactions at equilibrium states are called the standard potentials (abbreviated as, E^o). Under these conditions, the values of diverse redox systems are measured and cast in thermodynamic tables[2]. These potentials are determined against or versus (vs.) a reference system, which is often hydrogen electrode ($2H^+ + 2e^- \leftrightarrow H_2$) but other references are used as well. Note that the normal hydrogen electrode is given a potential of *zero volts* at the standard conditions (NHE, $2H^+ + 2e^- \leftrightarrow H_2$, $E^o = 0.00$ V).

The other mentioned reference systems have the following standard potentials: (SCE, $Hg_2Cl_{2(s)} + 2e^- \leftrightarrow 2Hg_{(l)} + 2Cl^-$, $E^0 = +0.268$ V), (SSCE, $AgCl_{(s)} + e^- \leftrightarrow Ag_{(s)} + Cl^-$, $E^0 = +0.222$V), and (CSC, $Cu^{2+} + 2e^- \leftrightarrow Cu$, $E^0 = 0.314$ V).

3. Redox potentials

The redox potential of a reaction refers to the tendency of the reaction to occur when compared to a reference system[1-8]. Redox potentials are often measured in aqueous solutions using electrolytic cells but organic solvents can be used for insoluble systems in aqueous media[9-10]. Normal hydrogen electrode (NHE) is usually employed as the standard reference system, which was assigned a standard potential of zero but other references, including calomel and mercury, are used as well. Keep in mind that redox potentials of reactions are always determined against reference systems, allowing to compare the ability of various systems to oxidize or reduce not just against reference systems but also between them as well. If a reaction has an oxidation potential superior to that of a reference system (e.g., NHE), this indicates that the reaction has more tendency to occur than the reference system. By contrast, if the reaction has a potential inferior to that of a reference system, the reaction will have less tendency to be driven towards the oxidized state but the reverse reaction will occur to form the reduced state.

In thermodynamics tables[2], the potentials of various redox systems (or couples) are often expressed in terms of reduction potential but the oxidation potential can be obtained by reversing the sign. For example, if the reduction potential of the redox couple (Ag^+/Ag) = 0.79 V vs. NHE, the oxidation potential can simply be written as: (Ag/Ag^+) = -0.79 V vs. NHE. Also, some thermodynamic tables listing the redox potentials are obtained in acidic media and others in alkaline media. Other aqueous media with variable pH values or even organic solvents could also be found in the literature[2]. Keep in mind that the potential values depend on the pH, concentration of the redox species, temperature (T), and pressure (P). These concepts will be discussed in later sections.

Although aqueous solvents are more often utilized in redox chemistry, non-aqueous solvents might sometime be employed when the redox species are insoluble in aqueous media. In this case, non-aqueous reference systems are required to sense the potentials of the redox couples dissolved in these non-aqueous media. The only problem is that non-aqueous reference electrodes suffer from long-term stability. The most utilized reference system suitable to sense the potentials in non-aqueous solvents is based on ferrocene (Fc^+/Fc). This redox couple is soluble in many organic solvents and its potential depends on the dissolving media. Examples include acetonitrile, dichloromethane, acetone, tetrahydrofuran, and dimethyl sulfoxide. To prevent significant fluctuations in potentials of non-aqueous reference systems, they should preferably be prepared fresh before each experiment. If compared to aqueous reference electrodes, the preparation procedure of non-aqueous reference systems is, fortunately, simpler and straightforward. For instance, a (Fc^+/Fc) reference electrode could simply be prepared by immersing an Ag/AgCl wire in a glass frit filled with a solution of acetonitrile containing ferrocene and an electrolyte (e.g. tert-buty chloride) at a specific concentration. The potential of this reference system should be around 400 mV vs. Ag/AgCl.

4. Redox potentials of elements across the periodic table

The knowledge of the electronic structures of elements across the periodic table allows the determination and interpretation of their chemical properties, including their oxidation and reduction tendencies and the number of electrons that could be donated or accepted. Keep in mind that redox potentials are measured quantities but the redox potential tendencies of elements across the periodic table could be guessed to some extent.

4.1. Tendencies across alkali metals group

All the alkali metals in the periodic table (Li, Na, K, Rb, Cs) possess an outer orbital s filled with 1 electron. Because this electron is farther from the nucleus and attached with minimum energy, it could easily be pulled out from the orbital. This results in lower ionization energies and modest electronegativities of alkali metals. In this group, the more the electron is distant from the nucleus, the more is it weakly attached to its nucleus. Thus, the electron can be ejected more easily in Cs than in Li. Therefore, alkali metal elements are very reactive and can easily donate their outer shell electron to more electronegative elements, such as O and Cl to form compounds, such as $CsCl_2$, NaOH, and NaH.

Consequently, alkali metals have relatively elevated oxidation potentials (in V vs. NHE): (Li/Li^+ = 3.05)> (Na/Na^+ = 2.71)< (K/K^+ = 2.92) = (Rb/Rb^+ = 2.92) = (Cs/Cs^+ = 2.92). Compounds susceptible to reduction could, thus, be reduced by alkali metals to gain 1 electron and become oxidized. A careful look on the changing trend of ionization energy and electronegativity values from Li to Cs and that of oxidation potentials reveal that Li has higher oxidation potential than the other alkali metals: Li/Li^+ > Na/Na^+ < K/K^+ = Rb/Rb^+ = Cs/Cs^+. This indicates that Li donates its electron more easily than the other alkali metals. The latter is attributed to the higher hydration properties of Li^+ in solution. Because of its smaller size, Li^+ has more tendency to hydration than other alkali metal cations. In turn, this increases its stability and hinders the removal of its outer shell electron.

4.2. Alkaline earth metals

Alkaline earth metals (Be, Mg, Ca, Sr, Ba) have 2 electrons in their outer shells. Therefore, they can lose one or both electrons at the same time when put in contact with oxidizing species. Alkaline earth metals are, thus, strong reducing agents but comparably less than alkali metals. The oxidation potentials of this group (in V vs. NHE) are: (Be/Be^{2+} = 1.85) < (Mg/Mg^{2+} = 2.37) < (Ca/Ca^{2+} = 2.76) < (Sr/Sr^{2+} = 2.89) < (Ba/Ba^{+2} = 2.9). Note that contrary to alkali metals, the trend of the oxidation potentials increases as expected from Be to Ba. The lowest oxidation potential value obtained for Be indicates its lower ability to donate electrons.

4.3. Transition metals

By contrast to alkali and alkaline earth metals in which participating electrons in the redox reactions come from only the s orbital, transition metals involve electrons from other orbitals, such as d. This is the reason why elements of this block are called the "d block". As a result, transition metals often have several oxidation states because they can lose or gain one or several electrons depending on the conditions. The most common oxidation states of transition metals are +2 or +3 but higher oxidation states up to +7 (Mn^{+7}) or + 8 (Os^{+8}) also exist.

The oxidation ability of transition metal atoms declines as the atomic number rises due to the increased attractive forces exerted by the nucleus on the outer shell electrons. For example, Zn with the outer shell electronic configuration of $d^{10}s^2$ can easily donate both electrons from the orbital s^2 to form Zn^{2+}. However, Cu ($d^{10}s^1$) can donate one electron from the orbital s^1 to form Cu^+ and another electron from the orbital d^{10} to yield Cu^{2+} even though the d orbital is fully saturated. The latter requires more energy and results in higher oxidation potential of the Cu/Cu^+ process.

4.4. Lanthanides and actinides

Redox processes with these elements involve electrons from the orbitals d, s and maybe f due to their higher atomic numbers, resulting in several oxidation states similar to the transition metals. The predominant oxidation state of lanthanides is 3+, when forming ionic salts. The electronic features of the actinide are similar to lanthanides. Lanthanides and actinides have typically elevated oxidation potentials ranging from 2 to 3 V. This results in their chemical activities, where they can easily react even by simple contact with air or water.

The properties of both actinide and lanthanide are summarized in Table 1 and 2 in terms of atomic number, electronic configuration of the outer shell, and relevant oxidations states.

Table 1: Some electronic properties of the lanthanides.

Element	La	Ce	Pr	Nd	Pm	Sm	Eu	Gd	Tb	Dy	Ho	Er	Tm	Yb	Lu
Atomic number (AN)	57	58	59	60	61	62	63	64	65	66	67	68	69	70	71
Outer shell electronic configuration (OSEC)	$5d^1$	$4f^15d^1$	$4f^3$	$4f^4$	$4f^5$	$4f^6$	$4f^7$	$4f^75d^1$	$4f^9$	$4f^{10}$	$4f^{11}$	$4f^{12}$	$4f^{13}$	$4f^{14}$	$4f^{14}5d^1$
Possible oxidation states (POS)	3	3, 4	3, 4	3, 4	3	2, 3	2, 3	3	3, 4	3, 4	3	3	2, 3	2, 3	3

Table 2: Some electronic properties of the actinides.

Element	Ac	Th	Pa	U	Np	Pu	Am	Cm	Bk	Cf	Es	Fm	Md	No	Lr
AN	89	90	91	92	93	94	95	96	97	98	99	100	101	102	103
OSEC	$6d^17s^2$	$5f^06d^27s^2$	$5f^26d^17s^2$	$5f^36d^17s^2$	$5f^46d^17s^2$	$5f^67s^2$	$5f^77s^2$	$5f^76d^17s^2$	$5f^96d^17s^2$	$5f^{10}7s^2$	$5f^{11}7s^2$	$5f^{12}7s^2$	$5f^{13}7s^2$	$5f^{14}7s^2$	$5f^{14}6d^17s^2$
POS	2,3	2,3,4	2,3,4,5	2,3,4,5,6	3,4,5,6,7	3,4,5,6,7,8	3,4,5,6,7,8	2,3,4,6	2,3,4	2,3,4	2,3,4	2,3	2,3	2,3	3

4.5. Other groups

The properties of elements can often be predicted according to their position in the periodic table because they follow similar tendencies within a group. However, the changing trend becomes more challenging starting from Group IIIA, representing the transition from metals and nonmetals. For example, in Group IVA, Sn and Pb are metals but Ge is semimetal. Elements of this group could yield several oxidation states, where C and Si often have +4, Ge may have +2 or +4, and Pb often has +2. In Group VA, P and N are nonmetals with dominant oxidation states of -3, +3 or +5. Sb and As are semimetals with a usual oxidation state of +3, and Bi is a metal that can donate up to 3 electrons to form Bi^{3+}. In Group VIA, O and S are nonmetals with higher electronegativities, often resulting in the negative oxidation state of -2. Te and Se are semimetals, with several oxidation states of −2, +2, +4, and +6. Finally, in Group VIIA, all the elements are halogens or nonmetals requiring only one extra electron to complete their outer electronic shell and yield the electronic configurations of noble gases. Thus, their most oxidation state is -1. However, in some compounds, Cl, Br and I may have oxidation states up +7.

Summary

The importance of reference systems (or electrodes) in redox reactions is highlighted. Reference electrodes allow determining redox potentials of species dissolved in either aqueous or non-aqueous solution. At the standard conditions (temperature 298 K or 25 °C, pressure 1 atm, concentration 1 mol L^{-1}), the potentials of redox reactions at equilibrium states are given by the standard potentials (abbreviated as, E^o). Examples of reference electrodes with their standard potentials in aqueous media are: (NHE, $2H^+ + 2e^- \leftrightarrow H_2$, $E^o = 0.00$ V), (SCE, $Hg_2Cl_{2(s)} + 2e^- \leftrightarrow 2Hg_{(l)} + 2Cl^-$, $E^0 = +0.268$ V), (SSCE, $AgCl_{(s)} + e^- \leftrightarrow Ag_{(s)} + Cl^-$, $E^0 = +0.222$V), and (CSC, $Cu^{2+} + 2e^- \leftrightarrow Cu$, $E^0 = 0.314$ V). In non-aqueous media, the most utilized reference electrode is based on ferrocene/ferrocenium redox couple (Fc/Fc$^+$), which is soluble in many organic solvents, such as acetonitrile. The potential of this reference system should be around 400 mV vs. Ag/AgCl. Note that redox potentials are measured quantities but the changing tendencies of elements across the periodic table can be guessed according to their outer shell electronic configurations. Except for Li$^+$, the oxidation potentials of alkali metals and alkaline earth metals increase within the group as the atomic number raise. For transition metals, their oxidation abilities decline as the atomic number rise due to the increased attractive forces exerted by the nucleus on the outer shell electrons. Lanthanides and actinides have typically elevated oxidation potentials ranging from 2 to 3 V, and the changing trend becomes more challenging starting from the Group IIIA, representing the transition from metals to nonmetals.

References

1. Ives, D. J. G.; Janz, G. J. (1961), Reference Electrodes, Theory and Practice (1st ed.), Academic Press.
2. Bard, A. J.; Faulkner, L. R. (2000), Electrochemical Methods: Fundamentals and Applications (2nd ed.), Wiley.
3. Zumdahl, S. S., Zumdahl, S. A. (2000), Chemistry (5th ed.), Houghton Mifflin Company.
4. Atkins, P.; Jones, L. (2005), Chemical Principles (3rd ed.), W.H. Freeman and Company.
5. Keith, O.; Myland, J.; Bond, A. (2011), Electrochemical Science and Technology: Fundamentals and Applications, Wiley.
6. Vladimir, S. B., Fundamentals of Electrochemistry, 2nd Edition, Wiley.
7. Dickerson, R. E.; Gray, H. B.; Haight, G. P. (1979), Chemical principles, (3rd edition), The Benjamin/Cummings Publishing Company, Inc., Menlo Park, CA.
8. IUPAC Definition of the Electrode Potential, Compendium of Chemical Terminology, (2nd ed.), (the "Gold Book"). Compiled by McNaught A. D.; Wilkinson, A. (1997), Blackwell Scientific Publications, Oxford.

9. Gritzner, G.; Kuta, J. (1984), Recommendations on Reporting Electrode Potentials in Nonaqueous Solvents, Pure Applied Chemistry, 56 (4): 461-466.

10. Pavlishchuk, V. V.; Addiso, A. W. (January 2000), Conversion Constants for Redox Potentials Measured versus Different Reference Electrodes in Acetonitrile Solutions at 25°C, Inorganica Chimica Acta. 298 (1): 97–102.

Section 4

Practical Questions and Problems with Solutions

A set of practical questions and problems with detailed solutions are provided to better understand the discussed concepts. The questions and problems range from simple to complex.

Q1. Briefly, define the reduction potential of a chemical species. At what conditions the standard reduction potentials are measured?

Ans1. The reduction potential is a measure of the tendency of chemical species to acquire electrons. The standard reduction potentials are measured at a temperature of 25°C, pressure of 1 Atm, and concentration of 1 mol L^{-1}.

Q2. Consider a hydrogen electrode with the standard potential (E^0 = 0 V) connected to a Cu electrode immersed in Cu^{2+} solution. i) Explain why electrons flow from the hydrogen electrode to the Cu electrode. ii) Write down the two half-redox reactions and the overall reaction. The standard potential of H$^+$/H$_2$ = 0 V vs. NHE and that of Cu^{+2}/Cu = 0.34 V vs. NHE.

Ans2. i) Electrons flow from one electrode to another if there is a difference in potential (or potential gradient) between the two electrodes. For instance, if two copper wires are connected together in acidic solution, electrons will not flow from one side to the other because the two electrodes are made of the same material and immersed in the solution. However, if a hydrogen electrode is connected to a copper wire immersed in Cu^{2+} solution, the electrons will flow from one pole to the other because of the difference in potential. The standard potential of H$^+$/H$_2$ = 0 V vs. NHE and that of Cu^{+2}/Cu = 0.34 V vs. NHE. Thus, copper has a higher potential or more affinity to gain electrons, which will reduce. By contrast, since hydrogen has a lower potential, it will oxidize or donate the electrons to Cu^{+2}.

ii) The two half-reactions could be expressed as follow:

Oxidation: H$_2$ → 2H$^+$ + 2e$^-$

Reduction: Cu^{2+} + 2e$^-$ → Cu

Overall reaction: H$_2$ + Cu^{2+} → 2H$^+$ + Cu

Q3. i) What standard reduction potential is assigned to the hydrogen electrode? Explain why? ii) Is the hydrogen electrode the only existing reference system? If not, give few other examples of reference systems. iii) Why are the other reference systems preferred over the hydrogen electrode?

Ans3. i) The hydrogen electrode is assigned a standard reduction potential of zero. It is used as a reference system to measure the potentials of other systems against it. ii) Other reference systems exist as well, including saturated calomel electrode (SCE) based on the reaction between elemental mercury and mercury(I) chloride (Hg$_2$Cl$_{2(s)}$ + 2e$^-$ ↔ 2Hg$_{(l)}$ + 2Cl$^-$), silver/silver chloride electrode (SSCE) involving redox reactions between silver metal (Ag) and its salt silver chloride ($AgCl_{(s)} + e^- \leftrightarrow Ag_{(s)} + Cl^-$), and copper–copper(II) sulfate electrode (**CSC**) implicating copper metal and its salt copper(II) sulfate ($Cu^{2+} + 2e^- \leftrightarrow Cu$).

iii) The other references systems are preferred over the hydrogen electrode because of the risky features of the hydrogen electrode. If hydrogen gas leaks out from the electrode and reacts with the oxygen present in air, an explosion will occur. The other systems are much safer to use.

Q4. i) What is the relationship between the reduction and oxidation potentials of any redox couple at the standard conditions? ii) If the oxidation potential of Na/Na^+ = +2.71 V vs. NHE, what is reduction potential of this redox couple?

Ans4. i) The reduction potential of a redox couple at the standard conditions differs from the oxidation potential in only the sign. If the reduction potential of a redox couple is E, the oxidation potential will have the value of $-E$. ii) If the oxidation potential Na/Na^+ = +2.71 V vs. NHE, the reduction potential Na^+/Na = -2.71 V vs. NHE.

Q5. Provide the values of concentration and temperature used for measuring the standard potentials.

Ans5. The values of the concentration and temperature used for measuring the standard potentials are 1 mol L^{-1} and 298 K (25 °C), respectively.

Q6. i) Which electrons are more susceptible to lose by transition metals? ii) How does the metallic behavior of transition metals vary with the oxidation number? iii) By which means could transition metals be stabilized at high oxidation states?

Ans6. I) Transition metals are likely to lose first the unpaired electrons present in their outer shells. ii) Transition metals with lower oxidation states will have more metallic behavior than those with higher oxidation states. iii) Transition metals at higher oxidation states can be stabilized by coordination with oxygen to form metal oxides. For example, Cr(VI) and Mn(VII) can form stable coordinated oxides, such as CrO_4^{2-} and MnO_4^-.

Q7. In the periodic table, how does the oxidation ability of transition metal cations (+2 or +3) change with the increase in atomic number?

Ans7. As the atomic number increases, the oxidation ability of transition metals cations (+2 or +3) generally decreases, reflecting the greater difficulty in removing electrons from transition elements with lower atomic numbers. The latter could be explained by the increased attractive forces exerted by the nucleus on the outer shell electrons of transition metals with lower atomic numbers. Transition elements with higher atomic numbers will have their outer shell electrons far from their nuclei, which make them easy to remove.

Q8. The oxidation potential of Al metal to Al^{3+} is 1.67 vs. NHE. i) Write down the oxidation reaction of Al in aqueous solution. ii) How many electrons are involved in this redox reaction? iii) What is the potential value to reduce Al^{3+} to Al? iv) Is Al^{3+} a good oxidizing agent?

Ans8. i) The oxidation reaction of Al to Al^{3+} can be written as: $Al \rightarrow Al^{3+} + 3e^-$

ii) This reaction involves 3 electrons, which are lost by Al metal during the oxidation process.

iii) The reduction potential of Al^{3+} into Al is -1.67 V vs. NEH. iv) Because the reduction potential of Al^{3+} into Al is too negative, Al^{3+} is not a good oxidizing agent.

Q9. Consider the elements of alkaline earth metals group: Be, Mg, Ca, Sr, and Ba. The respective electronegativities and atomic sizes of these elements are: (1.6, 1.3, 1.0, 0.95, and 0.89) and (0.89, 1.36, 1.74, 1.91, and 1.98). How would you expect the oxidation potential to vary along this group? Use the electronegativity and atomic size values to explain why.

Ans9. (Be, Mg, Ca, Sr, and Ba) belong to the second group of the periodic table called alkaline earth metals. They have 2 electrons in their outer orbitals, which could be lost in the presence of strong oxidizing agents to form cations with 2+ charges. The electronegativity is a measure of the tendency of an element to attract an electron in a chemical bond. The more the electronegativity is high, the more the element is hard to oxidize (or lose) an electron. Since the electronegativity decreases from Be to Ba, this means that Be has less tendency to lose electrons than Ba. In addition, the more the atomic radius is small, the stronger the outer shell electrons are attached to the nucleus, making it difficult to oxidize (or eject) the outer shell electrons. These two parameters could predict how the oxidation potentials should vary in this group. The oxidation potential or the ability to lose electrons should increase from Be to Ba.

Q10. Consider the following elements: K, Sc, V, Mn, and Co. i) Using the periodic table determine the electronic configuration of each element. Write down the outer shell orbitals in the form of $d^n s^m p^z$. Determine the oxidation states of each element and explain why. Gather all the answers in a table.

Ans10.

Element	Atomic number	Electronic configuration	$d^n s^m p^z$	Oxidation numbers	Explanation
K	19	$[Ar]4s^1$	$d^0 s^1 p^0$	K^+	K has one electron in the last s shell. Thus, it can only lose this electron to form a stable electronic configuration similar to that of Ar.
Sc	21	$[Ar]3d^1 4s^2$	$d^1 s^2 p^0$	Sc^{+3}	Sc can lose up to 3 electrons to form a stable structure

V	23	[Ar]$3d^3 4s^2$	$d^3 s^2 p^0$	V^{2+}, V^{3+}, V^{5+}	similar to that of Ar. Although other oxidation states may exist (+1 and +2), the most stable is 3+. V has a total of 5 electrons that could be lost through oxidation. In some cases, the two electrons of the s orbital will be lost to form V^{2+}. Once these electrons are gone, another one from the d orbital could be lost to yield V^{3+}. V can also lose all the outer shell electrons to form V^{5+}, such as in NH_4VO_3.
Mn	25	[Ar]$3d^5 4s^2$	$d^5 s^2 p^0$	Mn^0, Mn^{+1}, Mn^{+2}, Mn^{+3}, Mn^{+4}, Mn^{+6}, and Mn^{+7}	Mn has to total of 7 electrons in its outer s and d orbitals. Thus, it can donate up to 7 electrons passing through 1, 2, 3, 4, and 6 electrons. Examples of Mn compounds with these oxidation states are: $Mn_2(CO)_{10}$ (with 0), $MnC_5H_4CH_3(CO)_3$ (with +1), ($MnCl_2$, $MnCO_3$, MnO) (with +2), (MnF_3, $Mn(OAc)_3$, Mn_2O_3) (with +3), (K_3MnO_4 (with +5), (K_2MnO_4) (with +6), ($KMnO_4$, Mn_2O_7) (with +7).
Co	27	[Ar]$3d^7 4s^2$	$d^7 s^2 p^0$	Co^{-3}, Co^{-1}, Co^{+1}, Co^{+2}, Co^{+3}, Co^{+4}, Co^{+5}	Because the outer orbital s is full and that of d requires 3 electrons to saturate, Co can also gain up to 3 electrons to complete the d orbital. Meanwhile, it can donate 1, 2, 3, 4, or 5 electrons.

Q11. i) What types of electrodes could be combined with anodes and cathodes without influencing the overall cell potential? Give few examples of these electrodes. ii) Why is the hydrogen electrode not advised for practical use?

Q11. i) References electrodes (e.g., hydrogen, mercury, silver/silver chloride) could be combined with both anodes and cathodes. They can sense the potential of both anode and cathode without affecting the overall voltage. ii) For practical applications, reference electrodes other than hydrogen (e.g., mercury, silver/silver chloride) are used because of the potential risks of the hydrogen electrode. If hydrogen gas leaks out of the electrode and reacts with oxygen present in air, it will result in an explosion.

Q12. What is the standard potential of the normal hydrogen electrode? For what purpose is it used?

Ans12. The normal hydrogen electrode is assigned a potential of 0 V. It is used as a reference system to measure the potential of unknown systems against it. It is also employed to predict the feasibility of redox reactions in electrochemical cells.

Section 5

Electrochemical Cells and Redox Equilibria

Basic Concepts with Resolved Questions and Problems

Abstract

After highlighting the importance of reference systems (or electrodes) in the determination of redox potentials (Section 4), further details regarding their measurements are given in this section. These include the cell composition in terms of electrode and electrolyte and the main mathematical quantities governing the redox reactions at equilibrium states.

1. **Electrochemical cells**

The previously mentioned redox potentials are measured in electrochemical cells, typically made of electrodes and electrolytes assembled in two half-cells where the oxidation and reduction half-reactions occur separately[1-4]. Each half-cell consists of an electrode in contact with an electrolyte. The electrode might simply be a piece of conducting metal (e.g., Pt, Au, Cu, Zn, C, Fe). In each half-cell, the electrode might play the role of the cathode (reduction) or anode (oxidation), depending on the electrode potential and the experimental conditions. If the metal electrode serves only for electron transfer (collecting and giving electrons) and does not take part in the redox process, it is called inert electrode. Examples include noble metals, such as Pt and Au. However, if the metal electrode takes part in the redox reaction, such as by dissolving during oxidation (e.g., Cu, Fe, Mg), it is called active electrode.

The electrolyte is another important part of electrochemical cells. Electrolytes are often made of salts dissolved in solutions, such NaCl, $MgSO_4$, and KNO_3. Other low solubility salts (e.g., $BaSO_4$, $PbSO_4$, AgCl) are also used as solid electrolytes. In general, dissolved salts are good conductors of electricity but not in the traditional conductivity occurring in solid metal conductors. The electrical conduction through a piece of metal (e.g., Fe, Cu, Ni) occurs through the movement of the electrons across the holes present in the crystal structure but metal ions still remain in place. In salt electrolyte solutions, positively charged cations migrate towards the negatively charged electrode and negatively charged anions move in the opposite direction towards the positively charged electrode. This induces some sort of flow of charge which conducts electricity. Therefore, electrolytes allow the conduction and transfer of flow of charge to maintain the electroneutrality of the overall cell.

In electrochemical cells, electrodes immersed in electrolytes ensure the occurrence of the redox reactions. The two half-cells may contain the same or different electrolytes. The two electrodes (cathode and anode) could be immersed in the same electrolyte or separated by a membrane or salt bridge, built to prevent short-circuiting and maintain the overall cell electroneutral. In all cases, electrons generated at one half-cell from oxidation move through the external circuit (metal conductor) to the other half-cell, where they will be consumed by the reduction half-reaction. Therefore, a negative charge will accumulate at one pole and a positive charge at the other pole. To maintain the overall cell electroneutral and allow a continuous flow of charge, charged ions in the electrolyte will move towards the electrodes of the opposite charge to neutralize the accumulated charges at both poles.

2. Redox equilibria

Remember that redox reactions are chemical processes but not all chemical reactions are redox. The overall redox reactions appear similar to chemical reactions since the involved electrons in the half-reactions are canceled in the overall process[1-6]. For example, an overall reaction ($A_2 + B \rightarrow 2A^+ + B^{2-}$) is actually a redox process, which could be split into reduction and oxidation half-reactions:

Oxidation: $A_2 \rightarrow 2A^+ + 2e^-$ (1)

Reduction: $B + 2e^- \rightarrow B^{2-}$ (2)

Overall reaction: $A_2 + B \rightarrow 2A^+ + B^{2-}$ (1+2)

In the two half-reactions, A_2 and B^{2-} are the oxidants and A^+ and B the reductants. From the chemical viewpoint, A_2 and B are the reactants and A^+ and B^{2-} the products. The overall reaction is mass and charge balanced, and should proceed from left to right until reaching equilibrium. To determine how much reactants are transformed into products and how much is left in the reactant side, the reaction quotient Q is often used to describe the balance of the reaction and it progression state at any time[5-6].

Q for the overall reaction (1+2) is defined as: $Q = \frac{(B^{-2})(A^+)^2}{(B)(A_2)}$

where (A_2), (B) and (A^+), (B^{2-}) represent the activities (or concentrations) of the reactant and products, respectively. Remember that the activities are used for real concentrated solutions and concentrations for diluted solutions that often behave as ideal. Also, the activity (or concentration) of solid components, species in excess and electrons is always 1 (by convention). At the equilibrium state, the reaction quotient Q is expressed by the equilibrium constant ($Q = K_{eq}$).

Thermodynamically speaking, chemical reactions may occur spontaneously to acquire more stable states without requiring external energy. However, other reactions might require energy from external sources to move forward. At constant pressure and temperature, the Gibbs free energy (ΔG) is often employed to evaluate the spontaneity of a system. The process proceeds spontaneously if $\Delta G < 0$ and require external energy to occur if $\Delta G > 0$. $\Delta G = 0$ indicates that the system reached the equilibrium state.

Most electrochemical quantities, including electrode and cell potentials, are given at the standard conditions (298 K, 1 Atm, 1 mol L^{-1}). The standard free energy ΔG allows to calculate quantities at conditions different from the standard. The overall expression for the Gibbs free energy as a function of the reaction quotient is given by: $\Delta G = \Delta G^o + RT \ln Q$, where G is the Gibbs free energy, G^o is the standard Gibbs free energy, R is the gas constant (8.3144598(48) J mol^{-1} K^{-1}), T is temperature (in K), and Q is the reaction quotient.

In turn, G is linked to the electrochemical potential E by the expression: $\Delta G = -nFE$, where n is the number of the electrons transferred during the redox reaction and F is the Faraday constant (F = 96485 C mol^{-1}).

The replacement of G by E in the overall expression of Gibbs yields what is known as the Nernst equation: $E = E^o - \frac{RT}{nF} \ln Q$, where E^o is the standard potential. The Nernst equation is often streamlined by restricting the discussion to T = 25 °C, and by inserting the constants R and F by their values, the relationship becomes: $E = E^o - \frac{0.0257}{n} \ln Q$ or $E = E^o - \frac{0.0592}{n} \log Q$

Note the difference between the natural logarithm (Ln) and base-10 logarithm (Log). n represents the number of moles of electrons transferred during the process by taking into account the reaction stoichiometry.

At the equilibrium, $Q = K_{eq}$. The relationship becomes: $E = E^o - \frac{0.0592}{n} \log K_{eq}$

Keep in mind that the Nernst equation plays a very important role in redox chemistry for calculating quantities, such as concentrations of redox species undergoing oxidation or reduction, number of the electrons exchanged during the process, and potential of the electrodes or overall cell voltage.

3. **Spontaneity of overall redox reactions**

The potential of the overall cell reaction is calculated by summing the potentials of the two half-reactions as written in their oxidation and reduction forms: $E_{cell} = E_{ox}$ (anode) + E_{red} (cathode)

Keep in mind that the potential of each half-reaction is measured against a reference system, often the hydrogen electrode. The standard potentials listed in the thermodynamic tables are given either in the oxidation or reduction form, both equal in magnitude but have the opposite sign. In other words, if the oxidation potential of (A → A^+ + e^-) is E V vs. reference electrode, the reduction potential of (A^+ + e^- → A) is $-E$ V vs. reference electrode. These redox reactions and their potentials could also be expressed as redox couples: $E°(A/A^+) = - E°(A^+/A)$. Numerous examples are provided in the questions and problems section with detailed explanations as students often confuse whether potentials of the two half-reactions are summed or subtracted. All depends on how the two half-reactions are presented with their respective potentials.

The standard potentials listed in the thermodynamic tables vary between approximately -3 and +3 V. Thus, the maximum potential that an overall reaction could deliver is around 6V. This value induces a maximum equilibrium constant, $K_{eq} = 10^{100n}$ (obtained from the relationship: $6 V = \frac{0.0592}{n} Log\ K_{eq}$). This value is quite high when compared to equilibrium constants of other reactions, such as acid/base ($K_{eq}= 10^{-14}$). This means that the concentrations of the products at equilibrium issued from redox reactions are much higher than those of the reactants.

By considering the relationship between G and E, it can be concluded that overall redox reactions occur spontaneously if $E>0$, require external energy to proceed if $E<0$, and reached equilibrium state if $E = 0$.

Summary

The previous section focused on the importance of reference systems (or electrodes) in the determination of redox potentials at any given conditions. Here, the electrochemical cells used for measuring redox potentials are discussed in terms of electrode material and electrolyte. Electrochemical cells are basically made of two electrodes immersed in electrolytes. Two types of electrodes are mentioned: active electrodes taking part in redox reactions and passive electrodes only serving for electron transfer. Basically, the flow of charge in electrochemical cells begins at the anode where electrons are generated by oxidation. These electrons will then move through the external circuit (metal conductor) to the cathode, where they will be consumed by the reduction reaction. To maintain charge flow and prevent accumulation of negative charge at one pole and positive charge at the other pole, the charged ions of the electrolyte will move towards the electrodes of opposite charge for neutralization (electroneutrality). During redox processes, half-reactions (oxidation and reduction) occur at the electrodes, and their sum yields an overall cell reaction. At equilibrium, the reaction quotient of the overall process equals the equilibrium constant. The knowledge of the reaction quotient allows calculating the Gibbs free energy (ΔG), as well as determining the spontaneity of the reaction. In turn, ΔG is linked to the electrochemical potential E by the expression: $\Delta G = -nFE$. Cell potential is given by: $E_{cell} = E_{ox}$ (anode) + E_{red} (cathode). In sum, overall redox reactions occur spontaneously if $E>0$, require external energy to proceed if $E<0$, and at the equilibrium state if $E = 0$.

References

1. Bard, A. J. and Faulkner, L. R. (2001), Electrochemical Methods. Fundamentals and Applications, John Whiley&Sons, Inc, 2nd edition, USA.
2. Wahl (2005), A Short History of Electrochemistry, Galvanotechtnik, 96 (8):1820-1828.
3. Schüring, J.; Schulz, H. D.; Fischer, W. R.; Böttcher, J.; Duijnisveld, W. H. (1999), Redox: Fundamentals, Processes and Applications, Springer-Verlag, Heidelberg.
4. Atkins, A.; Jones, L. (2007). Chemical Principles: The Quest for Insight, W. H. Freeman.
5. Dickerson, R. E.; Gray, H. B.; Haight, G. P. (1979), Chemical Principles, 3rd edition, The Benjamin/Cummings Publishing Company, Inc., Menlo Park, CA.
6. Zumdahl, S.; Zumdahl, S. (2003). Chemistry (6th ed.), Houghton Mifflin.

Section 5

Practical Questions and Problems with Solutions

A set of practical questions and problems with detailed solutions are provided to better understand the discussed concepts. The questions and problems range from simple to complex.

Q1. i) Write down the mathematical expression linking the change in Gibbs free energy to the potential. ii) Define the Faraday number and provide its approximate value.

Ans1. i) The Gibbs free energy is linked to potential by the equation: $\Delta G = - nFE$, where n is the number of electrons exchanged during the redox process and F is the Faraday number.

ii) The Faraday number represents the charge of 1 mole of electrons and its approximate value is 96500 C mol^{-1}. In other words, 1 mole of electrons has a charge of 96500 Coulombs.

Q2. Consider an electrochemical cell composed of two terminals (cathode and anode). Write down the expression giving the difference in cell potential and Gibbs free energy. Estimate the change in Gibbs free energy if the cell generates a potential of 1.1 V and involves a transfer of 2 electrons. F = 96.5 kJ mol^{-1} V^{-1}

Ans2. The Gibbs free energy is linked to the potential by the equation: $\Delta G = - nFE$, where E represents the difference in potential between the two terminals (cathode and anode), F is the Faraday number, and n is the number of the exchanged electrons during the redox process.

The knowledge of the cell potential and number of electrons allows estimating the change in Gibbs free energy during the redox process.

For a cell involving 2 exchanged electrons and generating 1.1 V potential: $\Delta G = - nFE = - 2 \times 96.5 \times 1.1 = -212.3$ kJ mol^{-1}

Q3. If an electrochemical cell involving an exchange of 2 electrons undergoes a change in Gibbs free energy of -5.706 kJ mol^{-1}, what would be the difference in cell voltage? F = 96.5 kJ mol^{-1} V^{-1}. Could this cell be used in automotive starting engines or microelectronics?

Ans3. The Gibbs free energy is linked to potential by the equation: $\Delta G = - nFE$, where n is the number of the electrons exchanged, F is the Faraday number, and E is the cell voltage.

Thus, $E = -\frac{\Delta G}{nF} = -(\frac{-5.706 \text{ kJ mol}^{-1}}{2 \times 96.5 \text{ kJ mol}^{-1} V^{-1}}) = + 0.0296$ V

This cell delivers a very low voltage of only + 0.0296 V, which is far low to be considered for automotive starting engines. However, it could suffice to power microelectronics since they do not require high voltages.

Q4. i) Provide some examples of active and inert electrodes. ii) Why are inert electrodes preferred over their active counterparts for basic electrochemical measurements?

Ans4. i) Inert electrodes serve only for electron transfer and do not take part in the redox processes. Examples include noble metals (e.g., Pt, Au). By contrast, active electrodes participate in the redox reactions. Examples include Cu, Fe, Zn and Mg, which could dissolve to liberate metal cations and electrons during oxidation.

ii) The advantage of using inert electrodes for basic electrochemical measurements is that they limit the occurrence of secondary undesirable reactions. For example, the reduction of Fe^{+2} into metallic iron ($Fe^{+2} + e^- \rightarrow Fe^0$) on the Pt electrode will have no undesirable secondary reactions because Pt is hard to oxidize. However, if the same reaction is performed on non-noble electrodes (e.g., Cu), the deposited Fe on the electrode could be contaminated by traces of Cu resulting from the dissolution of Cu into Cu^{2+} ($Cu \rightarrow Cu^{2+} + 2e^-$) then its deposition along with Fe ($Fe^{+2} + e^- \rightarrow Fe^0$ and $Cu^{2+} + 2e^- \rightarrow Cu^0$).

Q5. Consider the redox couple Zn^{2+}/Zn with the standard potential of -0.76 V vs. NHE. Calculate the potential and free energy of this couple at 60 °C, 1 atm, and concentration of Zn^{2+} of 0.5 M.

Ans5. The reaction corresponding to the redox couple Zn^{2+}/Zn can be written as:

$Zn^{2+} + 2e^- \rightarrow Zn$

The Nernst equation can be used to estimate the potential at conditions other than the standard.

Accordingly, $E = E^o - \frac{RT}{nF} Ln\, Q = E^o - \frac{RT}{nF} Ln \frac{(Zn)}{(Zn^{2+})}$

Zn is a solid, hence its activity (or concentration) is 1, and the number of electrons involved in the reaction is n = 2.

This yields: $E = E^o - \frac{RT}{nF} Ln \frac{1}{(Zn^{2+})} = -0.76 - \frac{8.31 \times 333.15}{2 \times 96485} Ln \frac{1}{(0.5)} = -0.77$ V

It can be seen that although the concentration is lower than the standard (1 M), the resulting potential is slightly higher because of the elevated temperature.

The free energy could be estimated by the relationship: ΔG = -nFE = -2 × 96485 × (-0.77) = 148.58 kJ mol^{-1}

Note that the Gibbs free energy is positive, meaning that the reaction is not spontaneous under these circumstances but requires external intervention to deposit Zn^{2+} on the electrode. This could be done by applying a voltage of at least -0.77 V.

Q6. A student assembled an electrochemical cell using two half-cells. The first one is composed of a Pt wire immersed in Fe^{3+}/Fe^{2+} solution at the concentration of 0.01/0.02 mol L^{-1}. The other half-cell is made by another Pt wire immersed in MnO_4^-/Mn^{2+} solution at the concentration of 0.03 mol L^{-1} dissolved in acidic solution of 0.001 mol L^{-1}. The first half-cell is maintained at a temperature of 50°C and the second at 25 °C. i) Write down the reactions. ii) Calculate the overall cell voltage and change in Gibbs free energy. The standard potentials are: (Fe^{3+}/Fe^{2+}) = +0.77 V vs. NHE and (MnO_4^-/Mn^{2+}) = 1.49 V vs. NHE

Ans6. The comparison between the standard potentials of the redox couples indicates that MnO_4^- is more oxidizing than Fe^{3+} because of its high potential. Therefore, the reduction will occur at the Mn-based half-cell and oxidation at the Fe-based half-cell. This could also be confirmed by calculating the change in the oxidation states of the elements in both reactions. The oxidation state of Mn decreases from +7 in MnO_4^- to +2 in Mn^{2+}, ensuring the reduction half-reaction with the transfer of 5 electrons. The oxidation number of Fe increases from +2 in Fe^{2+} to +3 in Fe^{3+}, ensuring the oxidation half-reaction with the transfer of 1 electron.

Oxidation: $Fe^{2+} \rightarrow Fe^{3+} + 1e^-$

Reduction: $MnO_4^- + 5e^- \rightarrow Mn^{2+}$

The reduction reaction could be balanced by adding H^+ and H_2O to yield:

Oxidation: ($Fe^{2+} \rightarrow Fe^{3+} + e^-$) × 5

Reduction: $MnO_4^- + 8H^+ + 5e^- \rightarrow Mn^{2+} + 4H_2O$

To eliminate the number of electrons in the overall reaction, the oxidation half-reaction is multiplied by a factor of 5. The sum of the two half-reactions yields the following balanced reaction:

Overall reaction: $MnO_4^- + 8H^+ + 5Fe^{3+} \rightarrow Mn^{2+} + 4H_2O + 5Fe^{3+}$

ii) Since the experimental conditions of the two half-cells are different, it is better to calculate the potential of each half-cell using the Nernst equation before combing both.

$E_{Fe} = E^o - \frac{RT}{nF} Ln\ Q = E^o - \frac{RT}{5F} Ln\ \frac{(Fe^{3+})^5}{(Fe^{2+})^5} = -0.77 - \frac{8.31 \times 323.15}{5 \times 96485} Ln\ \frac{(0.01)^5}{(0.02)^5} = $ -0.77 + 0.0055 × 3.46 = -0.75 V

$E_{Mn} = E^o - \frac{RT}{nF} Ln\ Q = E^o - \frac{RT}{5F} Ln\ \frac{(Mn^{2+})}{(MnO_4^-)(H^+)^8} = 1.49 - \frac{8.31 \times 298}{5 \times 96485} Ln\ \frac{(0.3)}{(0.3)(0.001)^8} = $ 1.49 - 0.0051 × 55.26 = 1.20 V

The potential of the overall cell is obtained by summing both potentials: E_{cell} = 1.20 - 0.75 = 0.45 V

The Gibbs free energy is obtained using the relationship: ΔG = -nFE = -5 × 96485 × 0.45 = -217.09 kJ mol^{-1}

The value of Gibbs free energy is negative, meaning that the overall reaction occurs spontaneously under these conditions without requiring any external energetic intervention.

Q7. Calculate the equilibrium constant at the standard conditions of the following reaction: $AgCl \leftrightarrow Ag^+ + Cl^-$

The standard potentials of the involved redox couples are: (AgCl/Ag) = 0.2223 V vs. NHE and (Ag^+/Ag) = 0.799 V vs. NHE.

Ans7. The given redox couples would facilitate the determination of the two-half oxidation and reduction reactions. First, write down the redox couples in their given reaction forms.

$AgCl + e^- \rightarrow Ag + Cl^-$, $E° = 0.2223$ V (1)

$Ag^+ + e^- \rightarrow Ag$, $E° = 0.799$ V (2)

The comparison with the given overall reaction indicates that reaction (1) is the reduction and reversed of reaction (2) is the oxidation. Hence, the involved half-reactions in the overall reaction are:

Oxidation: $Ag \rightarrow Ag^+ + e^-$, $E° = -0.799$ V

Reduction: $AgCl + e^- \rightarrow Ag + Cl^-$, $E° = 0.2223$ V

Overall reaction: $AgCl \rightarrow Ag^+ + Cl^-$

Note that when the reaction is reversed, sign of the potential is also reversed.

The potential of the overall reaction at the standard conditions is the sum of potentials of both half-reactions written in their current states: $E°_{total} = -0.799 + 0.2223 = -0.577$ V

The equilibrium constant (K_{eq}) could be obtained by the Nernst equation: $E = E° - \frac{RT}{nF} Ln\ Q$

At the equilibrium, $Q = K_{eq}$ and $E = 0$. Therefore, $E° = \frac{RT}{nF} Ln\ K_{eq}$ or $Ln\ K_{eq} = E°(\frac{nF}{RT})$

The expression could also be converted to the 10-based logarithm (Log):

$Log\ K_{eq} = E°(\frac{n}{0.0592})$

The number of the electrons transferred during the process is n = 1.

$Log\ K_{eq} = (-0.577)(\frac{1}{0.0592}) = -9.75$, or $K_{eq} = 10^{-9.75} = 1.8 \times 10^{-10}$

Q8. Consider the overall reaction: $Zn_{(s)} + Cu^{2+}_{(aq)} \rightarrow Zn^{2+}_{(aq)} + Cu_{(s)}$, involving the two half-reactions.

$Cu^{2+}_{(aq)} + 2e^- \leftrightarrow Cu_{(s)}$, $E° = 0.34$ V

$Zn^{2+}_{(aq)} + 2e^- \leftrightarrow Zn_{(s)}$, $E° = -0.76$ V

Calculate the standard potential of the overall reaction. Is the reaction spontaneous, why?

Ans8. The overall reaction suggests that the first half-reaction is the reduction and the second is the oxidation. Therefore, the Zn-based reaction must be reserved to fit with the direction of the overall reaction.

Oxidation: $Zn_{(s)} \rightarrow Zn^{2+}_{(aq)} + 2e^-$, $E° = +0.76$ V

Reduction: $Cu^{2+}_{(aq)} + 2e^- \rightarrow Cu_{(s)}$, $E° = 0.34$ V

Overall reaction: $Zn_{(s)} + Cu^{2+}_{(aq)} \rightarrow Zn^{2+}_{(aq)} + Cu_{(s)}$

Note that the redox potential of Zn based half-reaction is reversed because the reaction is reversed to express an oxidation process.

The potential of the overall reaction is the sum of potentials of the two half-reactions:

$E_{Total} = 0.76 + 0.34 = 1.1$ V

Considering the relationship between ΔG and E ($\Delta G = -nFE$) and since E is positive, $\Delta G < 0$, meaning that the reaction is spontaneous.

Q9. Consider two half-cells with different concentrations of NaCl (1 M and 0.5 M). i) If both cells are separated by an ion membrane, what do you think will happen between the cells? ii) Does this setup create a difference in potential? iii) Suppose that instead of the semipermeable barrier, a permeable barrier is used then fully removed. What would happen in both cases?

Ans9. i) Because both half-cells have different concentrations, a concentration gradient will be created between them allowing ion circulation until reaching equilibrium. ii) This creates a chemical driving force and induces a difference in potential, which could be expressed by the Nernst equation.

$E = E° - \frac{RT}{nF} Ln\, Q$, where Q is the reaction quotient involving concentrations of the redox species.

iii) Semipermeable barriers are membranes allowing certain types of ions to pass through and block others. In this case, it allows the ions to diffuse slowly from one side (high concentration) to the other side (low concentration). The replacement of a semi-permeable barrier by a permeable one will allow most species to diffuse through and reach equilibrium faster. On the other hand, the removal of the barrier will induce an instant mixture of the ions in both half-cells, eliminating the concentration gradient and the difference in potential.

Q10. i) What is the relationship between cell voltage and free energy? ii) What is the link between the voltage and free energy of the overall cell and those of the two half-cells? iii) By comparing the potentials of the two half-cells, how could the anode and cathode be identified?

vi) What is the meaning of a positive overall cell voltage? v) What is the significance of a negative cell potential? vi) At equilibrium, what are the values of cell voltage and free energy?

Ans10. i) The cell voltage is related to free energy by the equation: $\Delta G = -nFE$

ii) The potential and free energy of the overall reaction is the sum of those of the two half-cells. In other words, these parameters are additives. iii) By comparing the potential of the two half-cells, it is possible to identify which poles are oxidized or reduced. Both potentials have to be compared at the reduction state. The half-cell with lower potential is the oxidation and the half-cell with higher potential is the reduction. This could also be confirmed or verified by determining the oxidation numbers of the elements.

iv) According to the above relationship, a positive cell potential means a negative free energy, implying a spontaneous reaction requiring no external intervention to occur. v) By contrast, a negative cell voltage implies a positive free energy. This, in turn, indicates that the reaction is not spontaneous and external intervention in terms of energy is required for the reaction to proceed. vi) At equilibrium, both the cell voltage and free energy equal to zero.

Q11. i) Estimate the voltage of a cell composed of two half-cells, each made of a Cu rod immersed in a solution of $CuCl_2$ at 3M. ii) What is the free energy of this cell at 20 °C? A student diluted the solution of one of the two half-cells with water until the concentration became 2 M. iii) Calculate the new overall cell potential and free energy. vi) From the practical viewpoint, is this cell worth constructing, why?

Ans11. i) Since the half-cells are made of the same metal immersed in the same solution, no voltage will be produced. ii) and i) The cell voltage is related to free energy by the equation: $\Delta G = -nFE$. Therefore, if $E = 0$, $\Delta G = 0$

iii) By diluting the concentration of one the two half-cells, a chemical driving force will be created due to the difference in concentration. This, in turn, will produce a potential, which can be estimated using the Nernst equation.

The potentials of both half-cells should first be estimated than compared. The lowest should be the oxidation and the highest the reduction.

At the oxidation state: $Cu \rightarrow Cu^{+2} + 2e^-$, and $E = E^o - \frac{RT}{nF} Ln(Cu^{2+})$

E_1 (3M) $= E^o - \frac{8.31 \times 383.15}{2 \times 96485} Ln(3) = 0.337 - 0.018 = 0.319$ V

E_2 (2M) $= E^o - \frac{8.31 \times 383.15}{2 \times 96485} Ln(2) = 0.337 - 0.0114 = 0.325$ V

The half-cell with superior concentration (or E_1) will be the oxidation and other pole with lower concentration (or E_2) will be the reduction.

Oxidation: $Cu \to Cu^{2+} + 2e^-$, E_1

Reduction: $Cu^{2+} + 2e^- \to Cu$, $-E_2$

Overall reaction: $Cu + Cu^{2+} (M_1) \to Cu^{2+} (M_2) + Cu$

This generates a cell potential: $E_{cell} = 0.319 - 0.325 = -0.006$ V

This potential is very small and negative in sign (non-spontaneous). Therefore, this cell is definitely not worth constructing.

The free energy of the overall cell can be estimated using the equation: $\Delta G = -nFE_{cell} = -1 \times 96485 \times (-0.006) = 0.578$ kJ mol^{-1}

The free energy is positive, meaning that the overall reaction is not spontaneous and requires external energy to occur.

Q12. Determine the two half-reactions involved in each of the following overall reactions. Calculate the potential of each overall reaction at the standard conditions. Which cells perform spontaneously?

i) $Fe^{2+} + Mg \to Fe + Mg^{2+}$

ii) $AgCl \to Ag + \frac{1}{2}Cl_2$

iii) $Cr + 3Cl_2 \to Cr^{3+} + 6Cl^-$

iv) $2Ag + \frac{1}{2}O_2 + 2H^+ \to 2Ag^+ + H_2O$

The standard potentials are: $(Mg^{2+}/Mg) = -2.37$ V vs. NHE, $(Fe^{2+}/Fe) = -0.44$ V vs. NHE, $(Cr^{3+}/Cr) = -0.74$ V vs. NHE, $(Cl_2/Cl^-) = +1.358$ V vs. NHE, $(Ag^+/Ag) = 0.7994$ V vs. NHE, and $(O_2/H_2O) = 1.23$ V vs. NHE.

Ans12. The overall reactions and the given redox couples allow an easy guess the two half-reactions in each case. The oxidation state (or number) method could also be used to determine the oxidation and reduction half-reaction in each case.

i) The oxidation state of Mg increases from 0 to +2 (oxidation half-reaction). The oxidation number of Fe decreases from +2 to 0 (reduction half-reaction).

Oxidation: $Mg \to Mg^{2+} + 2e^-$, $E° = 2.37$ V

Reduction: $Fe^{2+} + 2e^- \to Fe$, $E° = -0.44$ V

Overall reaction: $Mg + Fe^{2+} \to Mg^{2+} + Fe$

The Nernst equation could be used to calculate the potential.

$E = E^o - \frac{RT}{nF} \ln Q$, where Q is the reaction quotient linked to the concentrations (or activities).

At the standard conditions, since the concentration is 1M: $\frac{RT}{nF} \ln Q = 0$

Therefore, $E = E^o = 2.37 - 0.44 = 1.93$ V

The potential is linked to free energy by the relationship: $\Delta G = -nFE$, and since E is positive, ΔG is negative. This means that this reaction occurs spontaneously. The cell voltage is quite high, which makes it interesting for energy devices.

The same procedure is used for the other reactions.

ii) Oxidation: $Ag^+ + 1e^- \rightarrow Ag$, $E^o = 0.7994$ V

Reduction: $Cl^- \rightarrow \frac{1}{2}Cl_2 + 1e^-$, $E^o = -1.358$ V

Overall reaction: $AgCl \rightarrow Ag + \frac{1}{2}Cl_2$

$E = E^o = 0.7994 - 1.358 = -0.558$ V

Because E is negative, ΔG is positive, meaning that the reaction is not spontaneous but requires external energy to proceed.

iii) Oxidation: $Cr \rightarrow Cr^{3+} + 3e^-$, $E^o = 0.74$ V

Reduction: $3Cl_2 + 3e^- \rightarrow 6Cl^-$, $E^o = 1.358$ V

Overall reaction: $Cr + 3Cl_2 \rightarrow Cr^{3+} + 6Cl^-$

$E = E^o = 0.74 + 1.358 = 2.1$ V

The free energy is negative because E is positive, meaning that the reaction occurs spontaneously. The cell voltage is quite high, hence interesting for energy devices.

vi) Oxidation: $2Ag \rightarrow 2Ag^+ + 2e^-$, $E^o = -0.7994$ V

Reduction: $\frac{1}{2}O_2 + 2H^+ + 2e^- \rightarrow H_2O$, $E^o = 1.23$ V

Overall reaction: $2Ag + \frac{1}{2}O_2 + 2H^+ \rightarrow 2Ag^+ + H_2O$

$E = E^o = -0.7994 + 1.23 = 0.43$ V

ΔG is negative because E is positive, indicating a spontaneous reaction. The cell voltage is not that high but if several cells are combined in series, it could be useful for energy devices.

Q13. Consider the following overall reactions:
  i) $Sn + AgCl \rightarrow Sn^{2+}$ (0.07 M) $+ Ag + Cl^-$
  ii) $2Pb + 2SO_4^{2-} + PbO_2 \rightarrow 2PbSO_4 + Pb^{2+}$ (10^{-16} M)

Determine the oxidation and reduction half-reactions in each case. Write down the balanced reactions. Estimate the potentials of the half and overall reactions at 1 atm and 30 °C. Are the overall reactions spontaneous? The standard potentials are: (Sn^{2+}/Sn) = -0.137 V vs. NHE, (AgCl/Ag) = 0.222 V vs. NHE, (PbO_2/Pb^{2+}) = 1.468 V vs. NHE, and ($PbSO_4/Pb$) = -0.350 V vs. NHE.

Ans13. i) In the first case, the oxidation number of Sn increases from 0 to +2 (oxidation half-reaction) and that of Cl decreases from 0 to -1 (reduction half-reaction). This could also be confirmed by comparing the standard potentials. AgCl/Ag with higher potential should be the reduction half-reaction and Sn^{2+}/Sn with lower potential the oxidation half-reaction, which will generate the electrons. The two-half and overall balanced reactions are:

Oxidation: $Sn \rightarrow Sn^{2+} + 2e^-$, $E° = 0.137$ V

Reduction: $2AgCl + 2e^- \rightarrow 2Ag + 2Cl^-$, $E° = 0.222$ V

Overall reaction: $Sn + 2AgCl \rightarrow Sn^{2+} + 2Ag + 2Cl^-$

The potentials at these conditions could be estimated using the Nernst equation. Since no information is mentioned regarding the concentration of Cl^-, it has to be taken at the standard conditions (1 M).

$E_{Sn} = E° - \frac{RT}{nF} Ln\, Q = E° - \frac{RT}{nF} Ln\, \frac{(1)}{(Sn^{2+})} = -0.137 - \frac{8.31 \times 303.15}{2 \times 96485} Ln\, \frac{(1)}{(0.07)} = -0.137 - 0.0347 = -0.171$ V

$E_{AgCl} = E° - \frac{RT}{nF} Ln\, Q = E° - \frac{RT}{nF} Ln\, (Cl^-) = 0.222 - \frac{8.31 \times 303.15}{2 \times 96485} Ln\, (1) = 0.222 - 0 = 0.222$ V

The potential of the overall reaction is the sum of those of the two half-reactions written in their current states. $E_{cell} = 0.137 + 0.222 = 0.359$ V

$\Delta G = -nFE_{cell} = -2 \times 96485 \times 0.359 = -69.276$ kJ mol^{-1}

$\Delta G < 0$, meaning a spontaneous reaction.

ii) The oxidation state of Pb decreases from +4 in PbO_2 to +2 in Pb^{2+} (reduction half-reaction) and that of Pb increases from 0 in Pb to +2 in $PbSO_4$ (oxidation half-reaction). This could also be confirmed by comparing the potentials as in the previous case.

Oxidation: $Pb + SO_4^{2-} \rightarrow PbSO_4 + 2e^-$, $E° = 0.350$ V

Reduction: $PbO_2 + 4e^- \rightarrow pb^{2+}$, $E° = 1.468$ V

The reduction reaction could be balanced by adding H^+ and H_2O.

Oxidation: $(Pb + SO_4^{2-} \rightarrow PbSO_4 + 2e^-) \times 2$

Reduction: $PbO_2 + 4H^+ + 4e^- \rightarrow pb^{2+} + 2H_2O$

To eliminate the electrons in the overall reaction, the oxidation reaction is multiplied by a factor of 2.

The overall balanced reaction could be written as:

$2Pb + 2SO_4^{2-} + PbO_2 + 4H^+ \rightarrow 2PbSO_4 + Pb^{2+} + 2H_2O$

Or $2Pb + PbO_2 + 2H_2SO_4 \rightarrow 2PbSO_4 + Pb^{2+} + 2H_2O$

The potential could be estimated by the Nernst equation. Since nothing is mentioned about the concentration of protons and SO_2^-, they are taken as 1M.

$E_{PbO_2} = E^o - \frac{RT}{nF} Ln\, Q = E^o - \frac{RT}{nF} Ln\, \frac{(Pb^{2+})}{(H^+)^4} = 1.468 - \frac{8.31 \times 303.15}{4 \times 96485} Ln\,(10^{-16}) = 1.468 - (-0.24)$

$= 1.708$ V

$E_{PbSO_4} = E^o - \frac{RT}{nF} Ln\, Q = E^o - \frac{RT}{nF} Ln\, \frac{1}{(SO_4^{2-})^2} = -0.350 - \frac{8.31 \times 303.15}{4 \times 96485} Ln\,(1) = -0.350 - (0) = -0.350$ V

The potential of the overall reaction is obtained by summing both values: $E_{cell} = 0.350 + 1.468 = 1.818$ V

The cell voltage is positive, thus the free energy is negative, confirming the spontaneity of the overall reaction.

Q14. Consider the following balanced overall reaction: $Pt + O_3 + 2H^+ \rightarrow Pt^{2+} + O_2 + H_2O$

i) Calculate the standard potential of the redox couple O_3/O_2 if the standard potential of $Pt^{2+}/Pt = 1.188$ V and that of the overall reaction is 0.887 V. ii) What would be the cell voltage if the concentration of Pt^{2+} is diluted to 10^{-4} M?

Ans14. i) The overall reaction can be decomposed into two-half-reactions:

Oxidation: $Pt \rightarrow Pt^{2+} + 2e^-$, $E° = -1.188$V

Reduction: $O_3 + 2H^+ + 2e^- \rightarrow O_2 + H_2O$, $E° = ?$

Overall reaction: $Pt + O_3 + 2H^+ \rightarrow Pt^{2+} + O_2 + H_2O$, $E°_{cell} = 0.887$ V

Note that the sign of the potential of the Pt-based redox couple is reversed since it is written in the oxidation form. The overall potential is the sum of potentials of the two half-reactions written in their current states.

Using the Nernst equation, it is possible to estimate the overall potential.

$E°_{cell} = E°_{Pt} + E°_{O3}$, or $E°_{O3} = E°_{cell} - E°_{Pt} = 0.887 + 1.188 = 2.075$ V

ii) If the concentration is diluted, the new potential can be estimated using the Nernst equation.

$$E_{cell} = E^o - \frac{RT}{nF} Ln\, Q = E^o - \frac{RT}{nF} Ln\, \frac{(Pt^{2+})}{(H^+)^2} = 0.887 - \frac{8.31 \times 298.15}{2 \times 96485} Ln\,(10^{-4}) = 0.887 - (-0.118)$$
$$= 1.005\ V$$

Note that the concentrations of H_2O, O_2, O_3 and H^+ are all taken as 1 (excess, gas phase, no provided data).

Table of Content

Discount offers	1
Introduction	2
Section 4: Importance of Reference Systems in Determination of Redox Potentials	3
Abstract	4
1. Reference systems or electrodes	4
2. Standard potentials	4
3. Redox potentials	5
4. Redox potentials of elements across the periodic table	6
4.1. Tendencies across alkali metals group	6
4.2. Alkaline earth metals	7
4.3. Transition metals	7
4.4. Lanthanides and actinides	7
4.5. Other groups	8
Summary	8
References	9
Practical Questions/Problems with Solutions	11
Section 5: Electrochemical Cells and Redox Equilibria	16
Abstract	17
1. Electrochemical cells	17
2. Redox equilibria	18
3. Spontaneity of overall redox reactions	19
Summary	20
References	21
Practical Questions and Problems with Solutions	22
Table of content	34
About the author	36